Instrumental/Artificial Insemination of Honey Bee Queens

12 Most Frequently Asked Questions

Susan Cobey

Instrumental/Artificial Insemination of Honey Bee Queens
12 Most Frequently Asked Questions

Reprinted from The American Bee Journal from March 2016 Issue pg. 339-342

ISBN 978-1-908904-94-2

Published by Northern Bee Books, 2016
Scout Bottom Farm
Mytholmroyd
Hebden Bridge
HX7 5JS (UK)

Design and artwork
D&P Design and Print
Worcestershire

Introduction

As an introduction to instrumental insemination (I.I.) this article addresses the frequently asked questions about the technique. The answers are meant to give direction to further inquiryand to help evaluate the need and what is involved in mastering this skill. As a user and teacher of instrumental insemination, this article is in response to an increasing interest in this specialized technique.

Commercial queen producers recognizethe need for more rigorous programs to select, improve and maintain their breeding stocks. The public awareness of "CCD" and the movement of Africanized honey bees adds to this urgency. The development of micro-breeders and programs to select locally adapted and survival stocksare becoming more prevalent. Given these issues and concerns, a method of controlled mating is essential in achieving the goals of selective breeding.

Varroa is the key player in the challenge to keep our European honey bees, *Apis mellifera*, healthy and productive. This ectoparasite, feeds on the hemolymph of developing bees and is a major vector of pathogens. *Varroa* is a species complex. The shift of *Varroa jacobsoni* from its natural host, the Asian honey bee *Apis cerana*, to the Western honey bee, *Apis mellifera*, happened when European bees were introduced there. The mite jumped hosts, adapted and evolved to infest *Apis mellifera* requiring the re-naming of this parasite. The name given, *Varroa. destructor*, reflects its devastating impact on beekeeping.

In our 30-year history with *Varroa* in the U.S., infestations have traditionally been controlled chemically. The reduced efficacy of our arsenal of in–hive miticide treatments have turned attention to alternative controls and selective breeding.

The interest in selection of honey bee stocks that can manage, tolerate and/or show resistance to *Varroa* mite infestations has greatly increased. Of concern are the multiple pesticide residues in colonies resulting from exposure to miticides, and their compounding synergisticimpact in combination with agricultural chemicals residues.

The impact and movement of Africanized honey bees (AHB) is another motivator to establish bee breeding programs. In the southern U.S., the AHB is increasingly interbreeding with our domestic Europeanhoney bees. The dominating traits of the AHB, their defensive and migratory behaviors, are problematic. This is especially critical in providing pollination services. The AHB also poses a public health risk. The need to control and maintain breeding stocks to avoid the "Africanization" of our domestic European stocks is well recognized.

Stock improvement requires controlled mating and a program of systematic and scientific breeding methods. Instrumental insemination offers an essential tool to control honey bee mating. To gain the benefit of this tool also requires advanced beekeeping skills, the resources to run a selection and breeding program, and a long-term commitment.

Of critical importance to consider and research is the supporting information andmaterial on bee breeding methods. In the selection process, to maintain low *Varroa* mite levels and also maintain productive colonies involves selection for a complex of behaviors and traits. To ensure colony fitness, selection for a variety of traits such as productivity, temperament, overwintering ability, etc. must also be considered. The selection criteria must be weighted and balanced in terms of colony productivity and the "colony cost" of selection for "mite tolerance/resistance".

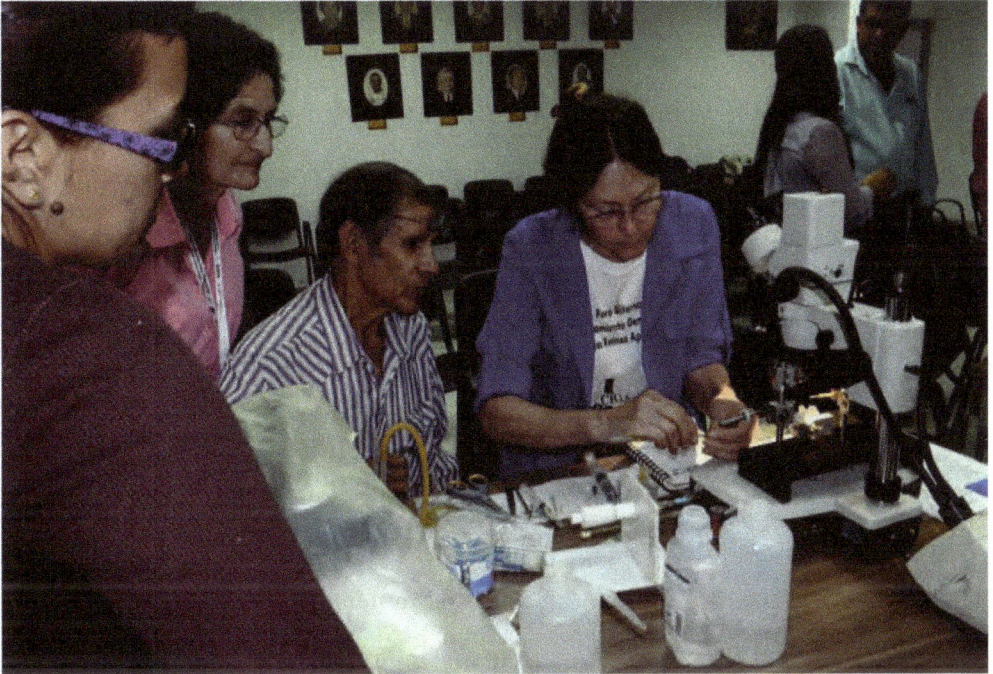

Susan Cobey teaching an Insemination Class in Tlapacoyan, Veracruz, Mexico Dec. 2015. Classes are also held annually in WA. state. *Photo by IAZ Martha Salazar Ulloa, UNAM*

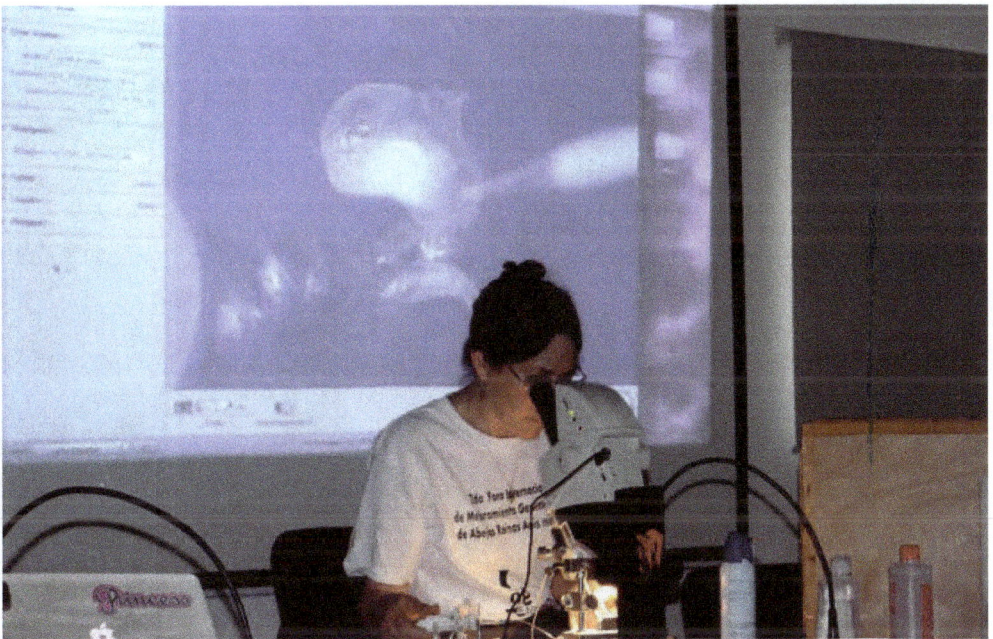

Susan Cobey demonstrating collection of honey bee semen. Tlapacoyan, Vera-cruz, Mexico Dec. 2015. *Photo by IAZ Martha Salazar Ulloa, UNAM*

Honey bees are unique, in that selection is based upon behavioral traits of a super organism with a complex social structure, in aconstant state of change. *Varroa* and its associated pathogens are also undergoing constant change. For these reasons, honey bee breeding can be challenging. Yet, with the current focus on bee breeding and advances of research and new scientific technologies, applying this knowledge in the field will hopefully offer real solutions. To gain the benefits of I.I., beekeepers must first build a solid foundation in establishing a breedingprogram or research project. Instrumental insemination is simply a tool in this process.

1. Instrumental Or Artificial Insemination, What's the difference?

The terms, artificial insemination and instrumental insemination, are used interchangeably. The term "instrumental insemination" was coined by Dr. Lloyd Watson, the first to successfully demonstrate a technique of instrumental insemination in 1926.

Eversion of the drone endophallus for semen collection, Susan Cobey and Luz Maria Saldaña. Tlapacoyan, Veracruz, Mexico Dec. 2015. Photo *by IAZ Martha Salazar Ulloa, UNAM*

He disliked the term artificial. The term artificial insemination is more commonly used and recognized by other industries such as; cattle, poultry, sheep, swine, equine etc.

2. Why Use Instrumental Insemination?

Instrumental insemination is simply a tool to control breeding. It also provides a means to create novel crosses for research purposes. Honey bee mating behavior is highly random and difficult to control. Queens mate in flight with numerous drones, averaging 15 to 20. Virgin queens fly to drone congregating areas, consisting of 10,000 to 30,000 drones from diverse genetic sources. The queen, who only mates during the first week or two of her adult life, stores the sperm collectedin her spermatheca for use over her lifetime. The queen stores only a small percentage of sperm from each drone she mated with and this is mixed in her spermatheca. All the drones the queen mated with are represented in the many subfamilies of her colony, although the ratio of these may change overtime.

3. Who Needs It?

I.I. is simply a tool for the bee breeder and researcher requiring specific crosses. It provides a method to control honey beemating. This technique enables the control of who the queen mates with, the number of drones she mates with and the semen dosage given and stored in her spermatheca.

Just because a queen is instrumentally inseminated does not mean she will be superior. The goal of producing Top Tier and"Rock star" breeder queens is dependent upon the selection of stock and the breeding program employed. In this process, many queens must be culled.

4. Is it Difficult to Learn?

Learning the technique requires; goodinstruction, precision equipment, practice, fine motor skills, good hygiene, patience and a commitment. The process can initially be awkward and frustrating to learn, as meticulousness precision is essential. The beekeeping knowledge and skills to rear aplentiful supply of mature drones and virgin queens is also required. To gain proficiency; practice, manual dexterity, and attention to detail are necessary. The pre and post care of queens also plays a critical role in the success of I.I.

To gain confidence in the technique I suggest, inseminating 30 queens in the initial practice. Keep good notes as to the success of each insemination. Bank the queens in a nursery colony for 2 days to allow for sperm migration. Check for vigor and mortality. Lethargic or dead queens indicate problems with injury or infection. Dissect the spermathecas of these queens to determine the rate of success. The color of the spermatheca will indicate how well inseminated the queen is, as described in reference listed: Dissection of the Spermatheca. Inseminate a second batch of queens and introduce these into nucleus colonies to establish them as laying queens.

5. What Equipment Is Needed?

The insemination instruments are specialized and vary widely in quality and price. The technique requires precision and accuracy in fine adjustments and this will determine the ease and repeatability of theprocedure. There is no standardization in equipment and some parts are not interchangeable between instruments. Therefore, research the options.

The basic instrument consists of a stand, a set of hooks or forceps, queen holder assembly, syringe and syringe tips. A microscope with a magnification of 10X to

20X, cold light and compatible stand with sufficient depth of field and instrument clearance are required. A cold light prevents heating and drying of tissues, a gooseneck L.E.D. light works well. A source of carbon dioxide, with a flow regulator and flexible tubing to the instrument, is used to anesthetize the queen during the procedure.

Modern instruments offer micro-manipulators that provide fine precision in movements. Large capacity syringes provide efficiency in semen collection, storage andshipment of semen. Various designs of sting manipulation tools offer personal choice in techniques. The Schley Instrument is currently the most widely used instrument, valued for its fine precision and wide range of flexibility in adjustments. The Harbo large-capacity syringe, designed for ease of semen handling and storage, is also popular and compatible with most instruments.

6. What Are The Most Common Problems?

Good sanitation and proper techniques arecritical to success. Injury and infection are the most common problems for beginners. Drones tend to defecate during the eversion process for semen collection, and can be very messy. Care must be given to maintain highly sanitary conditions and avoid feces contamination during the procedure.

The insemination procedure is very delicate and injury to the queen will produce poor results. Manipulations; opening and positioning the queen, bypassing the valvefold and insertion of semen, must be precise and brief. Queens vary physically,especially between the subspecies, and these nuances of differences must be learned and recognized.

Care of queens, their pre and post insemination treatment and introduction method, require careful attention. The queen signals her many changes of reproductive status to the workers, from virgin to mated and laying. These changes vary more among IIQs and therefore introductions need attentive treatment. These factors are discussed in the reference listed: Comparison of instrumental inseminated and naturally mated honey bee queens and factors affecting their performance, (Cobey 2007).

7. What Is The Most Overlooked Aspect?

Rearing a plentiful supply of select drones to maturity can be a major limiting factor. Queen production methods have been perfected and are routine. Most beekeepers are not accustomed to giving thesame detailed attention to drone production. Drones appear plentiful during peak season, yet are seasonally produced and the most vulnerable to stressors; such as parasites and pathogens, malnutrition, miticide and pesticide residues, poor weather conditions, etc. The colony will regulate the seasonal drone population based on many factors.

Strong healthy, well fed colonies are required for drone production. A colony can rear about 2000 drones during peak season, of which about half will be immature. Drones are mature at 2 weeks postemergence and peak at 3 weeks. Stressed colonies will eliminate mature drones and often continue to rear a new batch of young drones for the future. Colonies headed by older queens tend to rear and care for more drones. In extreme cases when seasonal conditions are unfavorable, queenless colonieswill hold and care for drones, although this method requires intensive management.

8. How Long Does The Procedure Take?

The timing required for the insemination procedure is dependent upon the skillof the inseminator, the quality of the equipment used and the quality and quantity of the live material produced. It is always best to overproduce queens and drones to ensure an adequate supply and allow for culling. In mastering this skill, it takes time to perfect techniques and gain proficiency. The beginner must learn proper procedures. Beekeeping skills essential to rear and care for virgin queens and drones are also essential.

The actual insemination procedure is very quick. Once the semen has been collected, it is a matter of seconds to insert the semen and inseminate the queen. The semen collection process is more time consuming and tedious. Timing is generally determined by the quality and maturity of drones. In general, about half the drones will yield usable semen. Many semen loads must be discarded due to contamination during the explosive force of the eversion. Some drones may not be mature and some simply will not yield semen.

A skilled inseminator can collect semen and inseminate 50 queens in 5 hours. The standard semen dosage per queen is 8 to 10 microliters. The procedure of semen collection and insemination can be separated. Collection of a 100 microliter tube of semen takes about 40 minutes. Each drone yields about one microliter of semen. Once the semen is collected, 40 - 50 queens can be inseminated in one hour. It is helpful to have a "runner" assist in supplying virgins to the table and caring for the inseminated queens after the procedure, including record keeping and marking queens.

9. What Is The Success Rate?

Success depends upon two major factors; the skill of the inseminator and the beekeeping skills to provide proper care of queens and drones. Assuming this expertise has been mastered, instrumentally Inseminated queens (IIQs) have the capability to preform as well or better than naturally mated queens, (NMQs). Better performance ofIIQs is based upon the ability to do selection and control the semen dosage given. The longevity and performance of the queen is largely based upon the genetic diversity of drones she mates with and how much semen she stores in her spermatheca.

The pre- and post-insemination caregiven to IIQs affects sperm migration and queen performance. The claim that IIQs do not preform as well as NMQs is unfounded. Tiring of this perception, I wrote a review of supporting studies. This discussion, listed in the references, (Cobey, 2007) clearly demonstrates how the differences in quality of care affect queen performance, over the actual procedure of insemination.

A frame of Carniolan drone brood in Washington, in April 2015. To rear a plentiful supply of drones to maturity from select colonies can be challenging.
Photo by S. Cobey.

10. How Long Can Honey Bee Semen BeHeld?

Honey bee semen can be held at room temperature for about 2 weeks and maintain good viability. The short-term storage of semen offers a huge advantage for the transport of semen and insemination scheduling. The type of diluent used and the temperature of storage (do not refrigerate) are important factors affecting viability, see (Cobey, S., Tarpy, D., Woyke, J. , 2013). New techniques are being developed for the long-term storage of bee semen at above freezing temperatures.

Advances in cryopreservation techniques for the long-term storage of honey bee semen now enable the conservation of select stocks and threatened subspecies, (Hopkins,B., Herr, C. Sheppard, W. 2012).

11

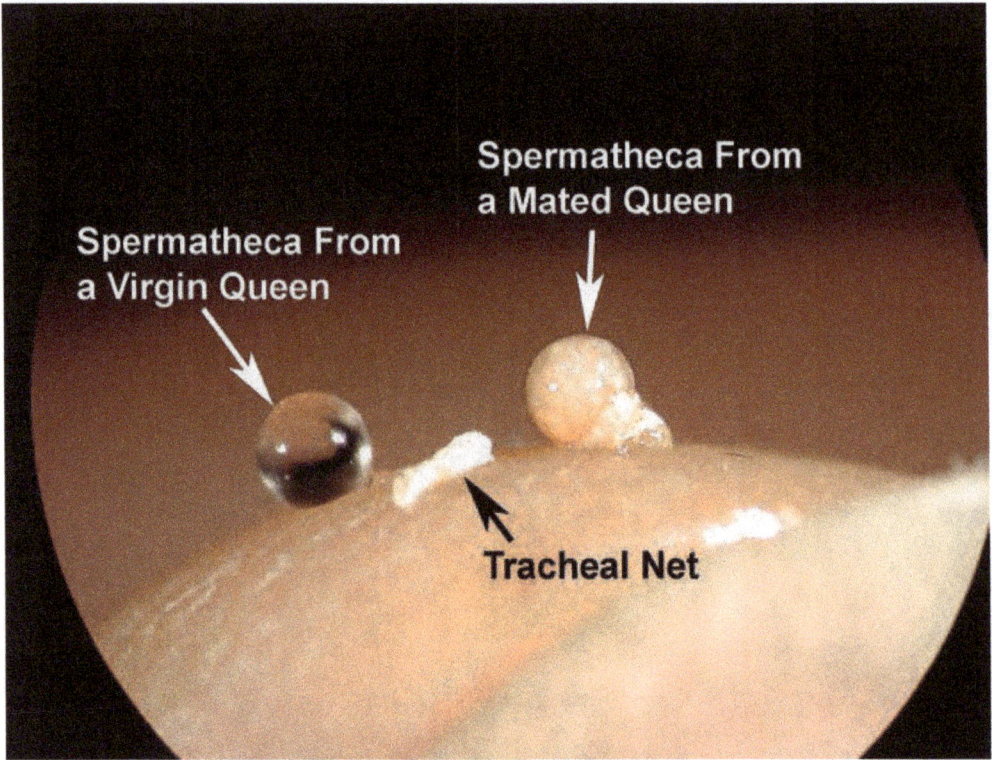

Color of the queen's spermatheca is checked to determine the success of insemination. The virgin queen's spermatheca is crystal clear and the mated queen's is a marbled "café au lait" color. Photo by S. Cobey.

Indigenous subspecies, threatened by the introduction of pests and pathogens and the introgression of imported non-native subspecies, can be preserved. This ability also offers unique breeding and research tools, as the ability to do selection across time.

Cryopreserved semen can be held indefinitely. Although, using current techniques there is some damage to the sperm. The laying patterns of queens inseminated with previously frozen semen are not sufficient to head productive colonies, although adequate to recover stock. Research is continuing in this area, with the worldwide interest in establishing honey bee germ-plasm repositories.

11. Can I Mix Semen From ManyDrones?

Genetic diversity increases colony fitness. Semen, from the numerous drones the queenmated, is naturally mixed and stored in the spermatheca. The ability to pool semen from a diversity of drone sources has advantages for breeding purposes and stock maintenance programs, although the techniques have not been perfected to date.

The natural mixing and migration of sperm from the queen's oviducts into the spermatheca is a complex process, requiring about 40 hours. This involves contraction of queen's muscles, mediated by the Bresslau's pump (the muscular system of the spermatheca duct) and specialized composition offluids in the semen and the oviducts, as well as active movement of the queen.

Perfecting a technique to homogenizing bee semen is difficult because the sperm tails are very long and fragile, and the semen is highly viscous, very dense and tightly coiled. Mixing techniques require dilution,mechanical movement and reconstitution of the semen, in which some critical components are removed. For routine inseminations, live active drones from different sources can be mixed in a flight box for semen collection.

12. What Can I.I. Do Beyond NaturalMating?

I.I. is a powerful tool for breeding as well as research purposes. It provides a means to create specific and novel crosses, beyond what occurs naturally. A single drone canbe mated to one or several queens, isolating and amplifying a specific trait. On the other extreme, semen from hundreds of drones can be homogenized and inseminated to a batch of queens. Varying degrees of inbreeding can be created to produce different relationships, including "selfing"; the mating of a queen to her own drones. Semen from the spermatheca of one queen can be extracted to inseminate another. These abilities provide a means to study and tease out the complexities of honey bees. The ability to store semen, over the short and long term, also offers many advantages.

Acknowledgements

Thanks to Photographer IAZ Martha Salazar Ulloa, Universidad Nacional Autónoma de México (UNAM)

Centro de Enseñanza, Investigación y Extensión en Ganadería Tropical Facultad de Medicina Veterinaria y Zootecnia (FMVZ)

Additional Information

Büchler, R., Andonov, S., Bienefeld, K., Costa, C., Hatjina,F., Kezic, N.,Kryger, P., Spivak, M., Uzunov,A.,Wilde, J. 2013. Standard methods for rearing and selection of *Apis mellifera* queens. In COLOSS BEEBOOK. Vol

1. http://www.coloss.org/coloss Specialissue. Jour Apic. Research 52.

Cobey, S. Insemination Fact Sheet Downloads: Eversion of the Drone. Semen Collection. Insemination of the Queen. Dissection of the Spermatheca.

www.honeybeeinsemination.com

Cobey, S.,2007. Comparison of instrumental inseminated and naturally mated honey bee queens and factors affecting their performance. A Review. Apidologie 38: 390-410.
www.honeybeeinsemination.com

Cobey, S., Sheppard, W. Tarpy, D. 2011.

Status of breeding practices and geneticdiversity in domestic honey bees. Chap. In: Honey Bee Colony Health: Challenges and Sustainable Solutions. Ed. Sammataro. D. & Yoder, J. Pp.25-36. CRC Press
www.honeybeeinsemination.com

Cobey, S., Tarpy, D., Woyke, J. 2013.

Standard Methods for Instrumental Insemination Of *Apis mellifera* Queens. In COLOSS BEEBOOK. Vol 1. http://www.coloss.org/coloss Special issue. Jour Apic. Research 52(4)

Cobey, S. 2015. Instrumental Insemination Techniques. Chap. In: Hive & Honey Bee. Dadant & Sons. Hamilton, IL. USA. Pg 791-801.

Hopkins,B., Herr, C. Sheppard, W. 2012.

Sequential generations of honey bee queens produced using cryopreservedsemen. Repro. Fertility & Developmt. Vol 24(8).

Koeniger, G., Koeniger, N., Ellis, J. Con-nor, L. 2014. Mating Biology of Honey Bees. Wicwas Press LLC. Kalamazoo, MI. USA

Page R.E., Laidlaw H.H. 1985. Closed Population Honey bee Breeding, Bee World 66, 63–72.

T. Rinderer, Ed. 1986. Reprinted 2010. Bee Genetics And Breeding. Academic Press.

www.ingramcontent.com/pod-product-compliance
Lightning Source LLC
Chambersburg PA
CBHW051433200326
41520CB00023B/7452